# The Fabric of Eternity

A Scientist's View of the Works of
Providence

## István Kolossváry

Edited by Kerri Lawnsby
Illustrations by Rita Farkas

The Fabric of Eternity:
A Scientist's View of the Works of Providence
István Kolossváry

Published by Biokol Books, Madison, New Jersey, U. S. A.

Cover design by Kerri Lawnsby
Cover art by © István Kolossváry
Illustration by © Rita Farkas

This book may be ordered directly from Amazon.com.

ISBN-13: 978-0988571709
ISBN-10: 0988571706

Printed in the United States

# Preamble

The *Fabric of Eternity* is my personal view of the Universe that allows for science and theology to explore the wonders of creation in peaceful unison. It is neither new science nor new theology; it is a humble attempt to lower the language barrier between these two disciplines in hope for a better understanding, appreciation, and gratitude to the Universe and her Creator.

I shall argue—on scientific grounds—that as far as science is concerned there are only two ways of thinking about the workings of the Universe. We either realize the continuous loving care of God, or we go it alone, reject God and insist upon a self-contained existence with no reason or purpose. Science can only offer these two fundamental alternatives but cannot decide between them in the lab.

I shall argue that the 'go it alone' theory may be proven false, though, in a scientific experiment involving *us*, humans.

Experiments are used to put scientific theories to the test whereby predictions of the theory are

∞

verified by a real-life device. If the device confirms the prediction, the theory may be right. A theory can be accepted with caution only if many different predictions are confirmed over and over and over again. However, if a single prediction is proven false in the experiment, the theory should be rejected no matter how clever and beautiful it might seem.

Some would argue that the most sensitive experimental device is a pencil standing on its tip; the tiniest push would make it tip over. I would argue, however, that the human person is even *more* sensitive and is an ideal instrument to experimentally verify the 'go it alone' theory. You can do the experiment yourself. I will show that the 'go it alone' theory makes predictions that contradict human nature. If we choose to go it alone we shall be bound — on scientific grounds — to deny free will, the idea of a unique person, being right or wrong, honor, human dignity, and the virtue of human efforts, just to mention a few of our commonly accepted traits. It is your call; does your experiment verify or refute the 'go it alone' theory?

∞

My own deep conviction vouches for a more humble Universe that humans can call home, a Universe that is alive by the love of God. To 'go it alone' means that we equate the Universe with *god*—the *god* of the philosophers, that is. The god of the philosophers is an absolute pure entity that is entirely self-contained, bearing its own cause, reason, and purpose. The god of the philosophers has no need for anything; it lives (if we can call it life) in perfect satisfaction and harmony with itself. God who loves the Universe, however, is beyond pure perfection and loves *us* human beings. It is love that lies beyond pure perfection; it is love that distinguishes *God* from the *god* of the philosophers. You know love only when you live it, it is beyond words.

Love does, nevertheless, have a link to science through the concept of altruism. Altruism means selfless care for the welfare of others exemplified by the likes of Mother Teresa on the pinnacle. But could there be any scientific discipline where selfless care for the welfare of others is part of the equation? Perhaps shockingly, the answer is *yes*.

The so-called Price equation inserts altruism in the paradigm of Darwinian selection where the only rule is the survival of the fittest.[1] The tragic life of George Price is a stark reminder, though, that beautiful mathematical theories sometimes just don't work. Price was a math genius but he realized when reflecting on his very own life that his theory was wrong. By way of logic the one irrefutable scientific and philosophical statement that can be made about altruism is that altruism is the only quality that can be added to pure perfection to enrich it in substance. This is the higher truth that forms the basis for anything we can reason about God; everything else is a matter of the heart.

I invite you to join me on my pilgrimage on the road from science towards theology as paved by Pierre Teilhard de Chardin in his life's work that culminated in *The Phenomenon of Man*. I have no claim of being right or wrong, and cannot promise we ever reach our destination, but... isn't the journey *itself* the destination?

---

[1] George R. Price, *Nature* **227** (Issue Nr. 5257), 1970, 520-521.

I am rooted in the Christian faith and tradition and you can hear my accent accordingly in everything I say. Therefore, where I felt necessary I give simple definitions to words that might be misunderstood if taken out of context. I want you to come along on this journey if you are interested; I love your accent no matter where you come from. I feel at a total loss for words to express my gratitude to every single person in my life who helped me prepare for my pilgrimage by heart and mind. Please join me.

István Kolossváry

November 2012
Madison, New Jersey

∞

∞

# Eternity Imprinted on a Tablet

*Eternity*—a heavy word and for all practical purposes it is understood as an ornate synonym for a time span stretching to infinity.

In Christian thought eternity takes a very different meaning. Simply put, eternity is the fullness of life; it is primarily a divine attribute. The following definition by Boethius is taken from Reginald Garrigou-Lagrange's seminal book, *Providence*:[2]

*Aeternitas est interminabilis vitae tota simul et perfecta possessio*

Translated:

*"Eternity is the simultaneous possession in all its perfection of endless life"*

Eternity is often likened to God watching the world from on high, at the summit of the

---

[2] Fr. Reginald Garrigou-Lagrange, *Providence. God's loving care for man and the need for confidence in Almighty God*, Tan Books and Publishers, Inc., Rockford, Illinois, republished in 1998.

proverbial mountain from where one can see the marching of generation after generation along their life paths, weaving the fabric of history. The foundation of this concept is that eternity is devoid of the dimension of time. God takes a timeless ownership of everything and anything that ever has been, is, and will be.

We humans, on the other hand, can only take ownership of our own lives to understand life as a sequence of event-beads threaded on an imaginary string we call time.

My first thesis considers how free will and providence interact, and points out a deep contradiction in the natural human awareness of the flow of time. The concepts of free will and providence are not exclusive to Christian belief; these concepts are well understood in a natural sense.

*Free will* means that we are capable of making decisions by considering conscious and unconscious factors pertaining to our decision. *Providence* refers to God's loving care and guidance throughout life.

∞

While most scientists accept free will, providence is scientific anathema. Even scientists who believe in a *god* of sorts, almost without exception flatly reject the idea that God has a personal interest in each and every one of us and, through *providence*, God offers direct guidance for our decisions in order to help us follow his divine plan.

Nonetheless, all scientists would agree (and maybe even take comfort) that there is a logical flaw with considering free will and providence in the context of a world where time flows. The flaw shows itself in a vicious circle which suggests that *our free will can change divine will*.

This concept is fundamentally contrary to Christian belief. Please, at this point, allow me to be overly simplistic so as to bring about the thesis. I am fully aware that my treatise calls for serious questions regarding predestination and all its synonyms (such as inevitability, karma, fate, and predetermination, to mention a few). Moreover, this notion raises questions about the presence of evil, not-so-happy endings, and suffering. Please refer to the Epilogue for my spiritual perception of the deeper meaning of providence.

Conceptualizing eternity as a mountain with people marching toward the summit evokes the hustle and bustle of history unfolding on the world stage. For each participant—whether a single person, a family, a community, a whole country, or even the whole of mankind—there is a path to follow on this stage, which has a beginning and an end. The beginning is determined, but the end depends on countless choices made at a great many crossroads. For each beginning God has a matching end in mind, to complete a journey with a divine purpose. Of course, as one of the most quoted sayings about God reminds us:

*His ways are not our ways*

We may only recognize the divine purpose when we have arrived at certain milestones in our life, or perhaps only at the end of our journey. Nonetheless, when at a crossroads, *providence* nudges the traveler toward this end, but it is *free will* picking the direction.

Again, here is the place for a disclaimer and a pointer to the Epilogue. The deeper meaning of providence dwells in the journey, *not* in the destination. It is about relationship with God, and not about arriving at a place of promise in this world.

To maintain the scientist's perspective I must start with a very simple model. Therefore, let us consider arriving at a fork along our journey's path at a particular point in time. At this point in time, we have a choice of continuing our way on different roads in different directions. *Providence* would hint to one of these choices and as long as we follow suit with our decision made of *free will*, logic and theology fit together. Thus, we travel along what we can call the *primary providential path* and in the end will find out the divine purpose of our journey.

However, if we take a different route, which is, again, our free will to make a decision, we leave the primary providential path and eventually may end up someplace else (to our demise).

We should not forget, though, that providence is infinitely resourceful and we are offered secondary, tertiary, and perhaps even more alternative paths to find our way home.

This is wonderful theology, but from a purely logical perspective, every choice we make to step off the primary providential path seems to provoke a *response* from God. It appears that by making a decision of my own free will I can change God's will and can force him to create new options. And, of course, I am not alone. Remember the cavalcade of history unfolding on the world stage. Countless decisions at every level throughout history bombard God to respond. God surely seems to be overwhelmed by the world. Nevertheless, of course, this argument does not stand to reason. God shall not be overwhelmed by the world and we cannot change God's will. There is nothing wrong with the awesome theology of divine providence; it is our concept of the flow of time that plays tricks on our thinking. A major revision of what we think about time is necessary to reconcile free will and providence.

Remember the sequence of event-beads threaded on an imaginary string we called time.

The key to the new thinking is that we throw away the string. Event-beads were snapshots of our lives and it was their unique sequence that caught our attention. However, as science has taught us, individual beads have their own reality. Beads can be arranged in a mathematical framework, which opens up a whole new vantage point for us to deepen our understanding. This mathematical framework took over 300 years to develop, primarily by René Descartes, Ludwig Boltzmann, David Hilbert, Hermann Minkowski, Albert Einstein, Erwin Schrödinger, and Julian Barbour.

The concept of a coordinate system is quite familiar. For example, you go to the stadium to watch a ball game. You enter through the main gate and check your ticket to find your seat. There will be three numbers that guide you; a section number, a row number, and a seat number. First, you find your section, then your row, and finally your seat. Or, you go to the opera. Again, three numbers; although in this scenario perhaps they are called different names; orchestra seat number 11 in row E, or seat 4 in the third row on the balcony. Three numbers place you at an exact location in our familiar, three-dimensional space. A particular system of mapping three numbers to a unique point in three-dimensional space constitutes a coordinate system. The most familiar coordinate system is called the Cartesian coordinate system where points are located on a three-dimensional grid spanning the three directions $x$, $y$, $z$ or left-right, front-back, and up-down using the more familiar labels.

Now, let us return to our seats and watch the action. Those of us who watch a ball game can see players running around in intricate patterns chasing a ball on the field. Those of us in the opera can also see movement on the stage; consider the grandiose movement of Swan Lake or La Traviata by Franco Zeffirelli. Of course, you can rightly argue that we go to the opera primarily for the music. Our focus for the moment, however, is on *movement* and besides, music itself has everything to do with movement, although a special kind of movement that vibrates in our ears.

A coordinate system gives us a convenient mathematical framework to decode still pictures. We know how to send the spectators to their seats, we know how to distribute an audience in the opera and let them be frozen in awe for the duration of the performance. Forget about clapping and cheering for now. But what about the playing field or stage? If we could freeze the action, give tickets with three numbers to the players or the singers and dancers telling them where to stand on the stage, we would have a good handle on the situation. But, to be honest with you, performers and audience staring at each other for a couple of hours wouldn't be too exciting, would it?

Here is another idea. We can give a deck of cards to each athlete or artist and tell them to move around following a pattern, which is encoded in their cards. Just go to the spot shown on the first card and, then, as you flip through the deck, keep moving where consecutive cards send you.

The exact same procedure could also be applied to move around the audience in their seats; although, with the exception of some ultra-modern theater productions, this wouldn't happen. The deck of cards method appears to have enriched the concept of a coordinate system to include movement, but in fact, it does not. At least, it doesn't to the extent that we need for *visualizing* providence.

This is because the *order* of the cards in the deck is *significant*. We have silently re-threaded the event-beads and re-introduced the flow of time; we are back at square one. We need something more, something qualitatively different from a coordinate system. The first step towards deeper understanding is capturing movement in an augmented coordinate system, which we shall call *composite space*.

Unlike coordinate space, which only considers positions, *composite space* considers positions and velocities/momenta together. The key to movement is velocity: the larger the velocity, the faster the movement. However, velocity is not simply speed; velocity also has direction.

A stroll on the beach along the Pacific Coast under the magnificent cliffs has a finite velocity component along the beach, but has zero velocity moving across or vertically. Skilled paragliders, on the other hand, can move along the beach while shooting up or swooping down in the air and also moving inland above the cliffs or gliding out to sea. Paragliders have non-zero velocities in all three directions.

In general, of course, everything that moves has velocities in all three directions. We can choose these directions to coincide with those defining positions; for example $x$, $y$, $z$ in a Cartesian framework or the more familiar perspective left and right, up and down, etc.

As long as position-directions and velocity-directions are the same, positions and velocities can be plotted *together* in the same framework we call composite space. Composite space has twice as many dimensions as the associated coordinate space. In the familiar three-dimensional world this number is equal to six: three spatial coordinates and three velocities.

This means that at any instant of time a moving object can be represented by a single point in composite space corresponding to its current position *and* velocity. Six-dimensional space is hard to visualize, but consider the case of walking along the beach, which involves only one direction. The associated composite space is now two-dimensional. We can plot the position of the walker along the beach on a horizontal line and the one-directional velocity on a vertical line. The two lines span a plane that can be visualized on a sheet of paper.

In this composite space, every *single* point represents the current state of motion of a moving object; its current position *and* its current velocity. The origin of this composite space where the position axis and velocity axis intersect corresponds to a person *standing* on the beach at a location that we arbitrarily assign the zero positional coordinate. This can be a beach-goer who just got out of his or her car, walked to the beach and now is standing there for a minute looking around to decide whether to start walking to the left or to the right. The scene

∞

may include other people walking by at the same zero-location. Their composite space points lie right on the vertical axis, above or below the horizontal axis depending on whether they pass by from the left or from the right.

Composite space can be thought of as a magic map where clusters of points light up showing the state of motion of all the people on the beach at any instant of time. As time flows by, the static set of points on the map comes alive like a swarm swirling along as a pictorial record of the collective stroll on a sunny afternoon at the beach.

Before making a final statement about composite space there is a tangential notion worth mentioning. When talking about the state of motion of an object, strictly speaking one should consider *momentum* instead of velocity. Momentum is simply velocity multiplied by the object's mass and reflects the simple fact that it does matter whether it is your cat jumping at you or a Husky. Henceforth I will stick with momentum, but it can always be interchanged with the more familiar velocity.

My final statement about composite space is that after a closer look it is very much like coordinate space with the same inherent contradiction between free will and providence. The proverbial mountain appears in composite space the same way as it does in coordinate space and the idea of the flow of time continues to shadow our thinking. The quantum leap in understanding time remains to be seen, but let not your heart be troubled, we have learned something utterly important. Unlike coordinate space envisioned to encode still pictures, composite space is conceived to encode motion.

The new key concept is that motion—position and momentum together—are captured in *still* images. As counterintuitive this may sound, take a look at Edgar Degas' equestrian sculptures; for example *Horse Trotting, the Feet Not Touching the Ground.*[3] These amazing works of art capture the magnificent trotting and galloping of horses with an astonishing vitality in a perfectly still, timeless bronze cast. If a

---

[3] See www.metmuseum.org/Collections/search-the-collections/120011343.

picture is worth a thousand words, some works of art speak volumes of science.

Our next step toward a deeper understanding of time and eternity requires a new level of abstraction. The idea works for both coordinate space and composite space leading up to the concepts of *configuration space* and *phase space,* respectively. The idea is very simple. You can think of it as multiply and conquer (instead of divide and conquer) to gain better footing in the world of abstraction. Think of a still life on a museum wall. It is a two-dimensional painting, which in fact depicts a three-dimensional scene involving multiple objects. The point here is that multiple objects are placed in the same abstract space. The abstract space can be an artistic rendering like the still life or as mundane as the now familiar coordinate space with points lit up where some objects are present. The next level of abstraction will require a new kind of coordinate space with a lot more dimensions. We are going to multiply the familiar number of dimensions (three) by the number of objects to be plotted in this new *configuration space*. Each coordinate axis

∞

of the configuration space is associated with only one (of the three) coordinates of a single object. In coordinate space, there is only one $x$ axis for all the $x$ coordinates of all the objects. Similarly, there is only one $y$ axis for all the $y$ coordinates, and one $z$ axis for all the $z$ coordinates in the Cartesian framework. Therefore, each object is represented by a single point in a three-dimensional space. In configuration space, however, considering $N$ objects (like $N$ people walking on the beach) there will be $N$ $x$ axes, $N$ $y$ axes and $N$ $z$ axes. The entirely new thing that these multiple axes allow us is going to be representing an entire scene with any number of objects as a *single* point in configuration space. Two different points will in general represent two entirely different scenes with multiple objects changing their place. Also note that every conceivable scene with different numbers of objects has its very own configuration space. Of course, there is a simple connection between any given configuration space and a coordinate space. Any scene-point in configuration space can always be dissected into its object-points in coordinate space.

∞

A visual example from my own research field would be a protein molecule. Protein molecules are the fundamental building blocks of key components of the living cell. These molecules themselves are built of smaller components called amino acid residues.

Protein molecules can be thought of as beads threaded on a string. The whole chain can wiggle, continuously changing its shape just like the wobbly spring toy Slinky.

In configuration space, Slinky is reduced to a single point lit up here and there like a lighting bug, depending on its current shape. The protein molecule is just a single point, too, its shape being encoded in the coordinates along the axes representing the locations of the individual beads.

When these coordinates are projected into three-dimensional coordinate space the whole protein molecule becomes visible to our three-dimensional vision.

Phase space is the twin brother of configuration space. Phase space is related to composite space in exactly the same way as configuration space is related to coordinate space. In phase space, points represent the entire motion-state of scenes comprised of multiple objects, their positions and their momenta. In addition to the $3 \mathrm{x} N$ coordinate axes, phase space also has $3 \mathrm{x} N$ momentum axes spanning altogether a $6 \mathrm{x} N$ dimensional abstract space. Now, before you get dizzy trying to visualize multidimensional abstract space and start considering whether to abandon ship, let me offer you a visual aid.

A point is a point in any number of dimensions. From what I have touched upon so far you have probably figured out already that the focus will be on points in space, not space itself. From now on (except when noted) space is a hand-drawn rectangle on a sheet of paper; you can think of it as a sandbox. We shall build whole worlds in this sandbox. Since the number of dimensions is irrelevant from our perspective we might as well utilize only two dimensions to visualize abstract worlds. We shall name this two-dimensional representation the *world tablet*.

Imagine the points of the world tablet through a magnifying glass. From a distance they look like ordinary points, but as you mentally zoom in, the points come to life. Depending on what kind of space the world tablet represents the points can mean very different things. Recalling some of the pictures used for illustration so far, the points can be entire scenes on the stage, whole groups of people walking on the beach, arrangements of still life themes, or, they can be your disorganized desk at the office, your living room, your small town, New York City, a whole ecosystem, the Earth with all of her inhabitants, or, a protein molecule, a living cell, a fish bowl, the ocean, a cloud, a hurricane, a single atom, or, yourself, a whole community or even the whole Universe.

The world tablet view helps us see the bigger picture. The most important thing about the world tablet is that *time is nowhere to be seen*. The world tablet is a collection bin of a great many different states of some reality that you can

visualize for yourself by zooming in on different world points on the world tablet. The exact meaning of what a *state* is will be paramount in our discussion later on.

For now, a simple word game will do. If the world tablet represents your brain activity, the world points correspond to your state of mind under different circumstances. Note, however that your state of mind is not considered in any timely fashion.

The concept of time is completely absent from this picture; or is it really? We shall see, but one thing is certain, time does not flow on the world tablet nor is time encoded in one of the dimensions as in Albert Einstein's theory of relativity.

Just a reminder, the absence of time does not mean the world tablet cannot capture motion even though in our everyday thinking time and motion are inseparable; just remember Edgar Degas' horses and phase space.

As a side note it is worth mentioning that in Julian Barbour's extreme view there is only one unique "state of mind" of the Universe that is real and as a logical consequence, motion is just an illusion and free will is a caprice.[4]

Later in this book I will address some of the extreme views of how far *rational* thinking in modern physics would go to exclude the supernatural.

Fanfares aside if you are still with me I have the honors to announce that we have reached a pivotal point of our journey to a better understanding of time and eternity in Christian thought. The new way of looking at our world beyond everyday thinking is through the concept of *phase space*.

Think of phase space in terms of the world tablet where every point represents the motion-state of an entire system of moving objects—the performance on the opera stage, the ball game on the play field, Degas' galloping horses, the

---

[4] Julian Barbour, *The End of Time. The Next Revolution in Physics*, Oxford University Press, Oxford, 1999.

wobbly protein molecule participating in the dance of life.

Now revisit the mountain where history is unfolding as God is watching from on high. This powerful image of eternity has set a logical landmine in our thinking that we are now in a position to defuse.

Let the world tablet display the fabric of history with every point representing a different snapshot of the marching generations in their upward spiral toward the summit of the proverbial mountain. The essence of the world tablet view is that we can look at a series of events *all at once*. We can look at event-beads <u>without threading them through the imaginary string of time</u> that many of us think is necessary to keep order in the world.

The idea of looking at different things all at once, though, may seem rather confusing. However, do not forget that the world tablet view is not about different things, *it is about the same thing in different states*. I admit, looking at different states of the same thing all at once may still seem dizzying, but hold on. Everybody is

∞

familiar with stereoscopic view. Technology comes in all shapes and sizes (and price tags) from the early day red and green glasses through the polarized goggles used in 3D movies, to the pinnacle of so-called immersion visualization where you are literally placed inside the 3D view utilizing a sophisticated projection system.

No matter what the technology, stereoscopic view comes from looking at an object from two slightly different angles at once. The result: a better, more detailed view enabled by depth perception. We can see things better in stereo. Better yet—holography. In a hologram you are looking at not only two, but numerous different views of an object resulting in an amazing 3D experience. In principle, whole movies can be encoded in a single hologram as if Degas' sculptures sprang to life. Of course, neither viewing in stereo nor watching a hologram can capture the idea in its entirety, but it helps appreciate the fact that looking at different states of the same subject all at once can indeed deepen our understanding of it rather than blur our vision.

In this perspective, God's view of the world is like looking at a mega-hologram. In the world tablet representation this means that the omniscient God can see all world points at once giving Him a full understanding of the whole world. God's view of the world is eternity:

*The simultaneous possession in all its perfection of endless life.*

Eternity is completely devoid of time, and while God is looking at the world as a whole, no detail is lost under his loving gaze.

Then, where are humans in this picture? We do not have a mega-holographic view of the world. We are localized to individual world points on the world tablet. We migrate from point to point with our whole environment. Ultimately, the entire Universe is migrating from world point to world point leaving a mark very much like an airplane leaves a tail on the sky. Every point on the world tablet is a potentiality, a possible state of the Universe. The world of potentialities is very, very lopsided. There is only a teeny-tiny fraction of world points that correspond to states where we can inhabit the Earth.

I shall come back to this point later. For now, let's only consider *actualities*; in other words, the path we follow on the world tablet.

On the dark screen of potentialities the actualities light up displaying a luminous path, which upon zooming in, reveals a record of history. One more time, note that depending on the scope of what you refer to as *the world* the history recorded in that light path can be as boring as the zigzagging of a tiny speck of dust, as exciting as our own life paths, or as grandiose as the history of mankind or the workings of the Universe. Since we have a free choice of the scope of the world let us personalize the analysis and put you on the light path. God looks at you through a mega-hologram being aware simultaneously of everything and anything that can possibly happen to you in your entire life time.

But what about you? You are localized at your own path that your life is drawing on your own world tablet. You can, or at least could, have full knowledge of the path you have already covered. You also have some ideas of what might have happened in the past or might happen in the future. Therefore, you are to some degree aware of potentialities slightly digressing from your actual life path. However, you could not possibly venture even a faint guess concerning the vast majority of all the potentialities of your life time.

Note that with regards to you, I used the words past, future and life time. Did somehow time creep back in our view? The answer is *yes* and *no*.

*No*: God's view of the world is absolutely timeless eternity.

*Yes*: In our limited existence in this world, held captive to our individual life paths, time does enter the picture. Time has something to do with the length of your life path on the world tablet, by some measure. In fact, with regards to phase space (where positions and momenta are considered together) time appears in a mathematical formula quantifying knowledge about certain properties of the objects in phase space. With perfect knowledge, time vanishes from the formula. The details are not important here.

What is important is that we have understood something deep about time and eternity and that will be the key to resolve the logical contradiction between free will and divine providence.

The logical land mine that was planted in our ordinary world view maimed our thinking and made it appear as though with our free willed decisions we could change God's will and prompt a response whenever we digressed from what we called the primary providential path.

Using the image of the cavalcade of history unfolding on the world stage, one could conclude that countless decisions at every level throughout history would constantly bombard God to respond and He would appear to be overwhelmed by the world.

But we have now arrived at the point where we can defuse this logical land mine once and for all by donning a new pair of holographic glasses so we can see the world stage in new ways. By displaying the world stage on the world tablet, the pieces fall into place.

It is highly simplistic, but it catches the essence to say that God never actually *re-acts* to our decisions; He already has the answer to any situation that can possibly happen. In eternity, He sees all things together as one whole system and in his infinite wisdom can gently assert His providential plan to the world throughout history in a single flash of eternal creation. We, on the other hand, are trapped in time, localized to our life paths and see the works of providence as an interactive play. We think that we ask all the questions and God must answer. What really happens according to Christian thought, instead, is that God has all the questions *and* all the answers. We shall see a bit later that this statement is more subtle, though. Right now our focus is on picturing providence and its coexistence with free will.

Take a final look at these concepts from your own perspective. Put yourself on your life path in the world tablet representation. The life path is a snaking curve—a trajectory. The trajectory has a starting point that was fixed when you were born, and a current ending point where you are in life right now. Your life story is told along the trajectory from the starting point to the current ending point. As your time is ticking along, the end point of the trajectory will move ahead on the world tablet by tiny installments as you keep extending your life path by making a long series of decisions. Each decision will select a new point in the vicinity of the *then* current end point. Everybody makes decisions so it goes without saying what is involved.

As we mentioned already you have information regarding your life not only from your actual past, but also from a limited set of potentialities from your past, presence and even the future. This watershed of information surrounds your trajectory on the world tablet like a halo. Every time you make a decision it is based on information gathered from this halo. Of course, many of your decisions will be entirely determined by the information at hand. Such decisions are not really decisions and can be conscious or unconscious. Nonetheless, most people will agree that free will, which is of course not unique to Christian thought, is real and you do have a choice at least at some crossroads along your life path. These free willed decisions determine where your trajectory on the world tablet will come to rest at the time of your passing.

What providence means is that not all ends are equal and that God guides you through your life journey toward an end that meets His providential plan. In other words, God fills your life with a purpose. Your life has a purpose. My life has a purpose as does everybody else's. Humankind has a purpose and so does the entire Universe. This is Teilhard de Chardin's Omega point: the purpose of evolution.[5]

Nobody in their right mind would call evolution into question, but Christian thought asserts that evolution is not totally random, with the sole

---

[5] Pierre Teilhard de Chardin, *The Phenomenon of Man*, Harper Perennial Modern Thought, New York, 2008.

pseudo-purpose of survival. Evolution as a whole has a purpose of reaching the Omega point where God's providential plan is fulfilled. No doubt, randomness has its full potential unleashed in exploring the realm of possibilities, but randomness is only a servant and not the master, of evolution.

You may wonder who would dare ask if we could possibly know anything about the Omega point? A question like that may sound foolish, but isn't it in our human nature to always ask the questions? If we didn't inquire how would we hope to ever get answers? We know the Omega point is out there, but the faint halo of what can be known to our feeble minds won't illuminate it. Yet there is something of utmost importance to be said about the Omega point which I discuss in the Epilogue. To continue unfolding the fabric of eternity it suffices to say that the Omega point is all about relationship; relationship between the almighty God and the limited human.

One aspect of this relationship and, the focus here, is providence. Providence comes in many shapes and forms on all levels of existence and always involves divine interaction with the created world. For many, if not most men and women of science, it is this very interaction between God and the Universe that is so unacceptable.

I'd like to simply show you that such interaction cannot be denied on scientific terms. In fact, denial inevitably leads to outlandish ideas that conform only to some advanced mathematical formulas that rob humans of our very human nature.

When we think of God the first thing that comes to mind is transcendence. God transcends the world and if we limit ourselves to the world what need do we have for God, even if we accept that God has created the Universe? This is how most scientists think. Of course God is transcendent, but God is also immanent; He is present in his creation. This is not some Christian dogma. This may well be a mystery, but can't we get a glimpse of this mystery when we discover the spirit of a master in his or her creation of a work of art, a scientific theory, a philosophical train of thought, even a molecule? Chemists working in the same lab can often identify the person who synthesized some molecule by recognizing the character of the person in the characteristics of that molecule. Arthur Peacocke has advocated the idea and coined the term "whole-part-influence" to suggest that the immanence of God is not only a passive imprint of his being in the world, but also an active way of exerting His providence.[6]

The immanent God constantly interacts with the world *as a whole* with the effect of the mosaic pieces of the world falling into place to create an image that could not be envisioned at the scale of the individual pieces. We might say that the world is more than the sum of its constituents, because God is immanent to the world.

Science has a similar paradigm called *emergent phenomena*. Emergent phenomena were first recognized in molecular systems where molecules showed evidence of self-organization

---

[6] Arthur Peacocke, *Paths from Science Towards God. The End of all our Exploring*, Oneworld Publications, Oxford, 2001.

that could not be predicted by the laws of physics applied to them individually. Such emergent behavior is attributed to the molecules being incorporated into a system as a whole; henceforth the term *whole-part-influence*. With recent advances in science, emergent phenomena may turn out to be more the rule than the exception, but most if not all scientists would argue that it is complexity that explains them, not an immanent God.

Whatever *complexity* means, it is nonetheless internal to the system and, ultimately, internal to the Universe. Whole-part-influence, however, goes beyond that. While internal to the Universe in its workings through immanence, whole-part-influence nevertheless originates in the transcendence of God. The world is *in God* as Scripture reminds us. The world is breathing God, one might say. Now I dare you: take a deep breath and plunge into the depths of the Quantum, reading on, to see more of the marvels of the fabric of eternity.

# Pondering the Quantum

The Quantum; veiled in mystery, weird, impossible, only for nerds, giving you and me the creeps. But in reality, the world of quantum physics is much closer to the human psyche than the machine-like macro world that we claim as known territory. Isn't the human psyche veiled in mystery, impossible, only for shrinks, and a source of great anxiety?

We shall set sail on the Quantum Ocean in a tiny raft, but like the ancient Greeks we'll never leave sight of the shore. The holographic view of the world tablet with its increasingly abstract spaces is well suited for discussion of quantum theory. The world tablet view embodies different *states* of a world; and *world* can be as tiny as a single subatomic particle or as grand as the whole Universe, and anything and everything in between including ourselves. Quantum physics deals directly with *states* and thereby has a natural link to the world tablet. Unlike human laws which require hundreds and thousands of

pages to capture, the law of the Quantum is captured in exactly 20 characters in one of the three most famous equations in the history of physics—the time dependent Schrödinger equation.[7]

The mystery of the Quantum can be attributed to a single property of Schrödinger's equation: *if two different states are allowed by the equation then any linear combination of them is also allowed.*

This so-called *superposition* principle has deep consequences and surely is difficult to reconcile if expressed in everyday terms. For example, if you toss a coin you fully expect it will be heads *or* tails. According to the superposition principle, the result of a coin toss can be heads or tails, heads *and* tails, or even *somewhat* heads and *mostly* tails or vice versa. Weird, isn't it?

Expand this notion to the expectation that a person cannot be in two different places at the same time. Popular murder mysteries often try

---

[7] Note that the most famous equations in the history of physics include Newton's equations of motion and Einstein's field equations of General Relativity.

to convince the reader otherwise but in the end it is always just a clever trick by Houdini's principle which is make the audience believe what you want them to believe.

With Quantum Theory, however, there is no magician's trick involved. Truly, things *can be* located at two or more different places at the same time, sometimes more here than there.

One of the seminal quantum experiments has shown countless times that the superposition principle is at work when a *single* photon, which is the indivisible basic unit of light, can in fact pass two adjacent slits in a wall at the exact same time. This and other experiments provide irrefutable evidence of the superposition principle.[8]

How, then, is it that we never see superposition happening in the macro world? This is the ultimate $64 Million question in quantum physics and is called *the measurement problem*.

---

[8] The best short story of the Quantum was written by John Polkinghorne; *Quantum Theory: A Very Short Introduction*, Oxford University Press, Oxford, New York, 2002.

I am certainly not in a position to answer this question, but the measurement problem provides the backdrop for *my* ultimate question: how do we inject whole-part-influence in the Universe?

First, however, I'd like to take a short scenic detour. Just for fun, let me introduce to you yet another level of abstraction into the series of coordinate space, composite space, phase space, and configuration space. Did you know that if you get dizzy in an IMAX theater you just close your eyes to stop your head spinning? So not to worry, if you get dizzy reading this chapter just remember the world tablet. The world tablet is the safety zone for an unhealthy dose of abstraction and it is all we need to think about the fabric of eternity.

The particular space where quantum states are considered is called Hilbert space, named after David Hilbert (the most famous mathematician of the turn of the twentieth century) who invented it before the quantum revolution began. Before Hilbert space, all spaces had axes with properties describing the motion of objects (position combined with momentum/velocity). In particular, configuration space was the venue for plotting states of an entire system of moving objects. There is an equivalent of configuration space in the quantum world but its axis properties are somewhat different because the motion of quantum objects is described by Schrödinger's equation, not Newton's. More importantly, however, configuration space has limited use because its inhabitants are single states, making it cumbersome to deal with the superposition principle.

The genius of David Hilbert was to make the states *themselves* span his space. The axes of Hilbert space represent multiples of all possible single states, or *pure states*, which are often infinite in number. Henceforth, naturally, every single point in Hilbert space plotted against the axes is a superposition of pure states.

Each axis can be viewed as a ruler with marks. The pure states that are sometimes amenable to our macro world are always situated at the '1' mark of the ruler. This is, of course, highly abstract but remember the safety zone: the

∞

world tablet encompasses Hilbert space while filtering out all of its complexities.

Notwithstanding the beauty of Hilbert space I wouldn't mention it if it weren't for a simple analogy that I want to show you on the scenic detour that follows. Deep down still highly abstract, yet inherently familiar in our everyday lives, are numbers; especially positive integers 1, 2, 3, and so on which are also called natural numbers. Such numbers come in two flavors: primes and composite numbers.

The ever elusive *primes* are numbers divisible only by one and by themselves. The smallest prime is 2 then comes 3, 5, 7, 11, 13, 17, etc. By definition 1 is not considered to be a prime. There are an infinite number of primes and finding big ones—*really big ones* with thousands of digits—has been a popular game for computer geeks.

The fundamental theorem of number theory asserts that *composite* numbers can always be written as the product of a unique set of primes. For example, the smallest composite number is 4 which is equal to 2 times 2; then 6=2*3, 8=2*2*2, 9=3*3, 10=2*5, and so on.

Primes are not only fascinating, they are indispensable in cryptography. The practical difficulty of factoring huge composite numbers (i.e., finding their prime components) is the basis of sending encrypted e-mail messages, credit card numbers and other sensitive information over the Internet.

From our current perspective, however, prime numbers form the basis of a better understanding of Hilbert space. Think of prime numbers as pure states and composite numbers as superposed states. We can construct a toy Hilbert space by means of a multidimensional rectangular grid (infinite in dimension) with axes representing multiples of the prime numbers. Consider the two dimensional example of a spreadsheet with rows and columns: rows correspond to multiples of 2 and columns correspond to multiples of 3. As a matter of convenience also include the multiple of zero; in other words, start numbering rows and columns with zero.

∞

This spreadsheet will then contain every number that can be written as a product of any combination of the primes 2 and 3. The upper left cell in row zero and column zero—henceforth referred to as r0c0 ( 'r' stands for rows and 'c' for columns)—contains the number 1 (neither 2 nor 3 among its factors).

We can fill the remaining cells as follows. r0c1=3, r0c2=3*3=9, r1c0=2, r1c1=2*3=6, r1c2=2*3*3=18, r2c0=2*2=4, r2c1=2*2*3=12, r2c2=2*2*3*3=36, and so forth.

Including the third prime number 5 can be used to extend the spreadsheet in three dimensions to plot: r2c3f1=2*2*3*3*3*5=540 or r0c1f2=3*5*5=75 where 'f 'stands for floor representing the third dimension.

Further extension into higher dimensions is straightforward. I hope you found this short excursion into the land of numbers helpful to familiarize the concept of Hilbert space. Now back to the measurement problem.

The measurement problem arises when we try to measure some property of a quantum object. We shall skip the technical details and focus only on the essence to shed light on how whole-part-influence could be injected in the Universe.

According to the superposition principle, quantum objects are like Janus the Roman god with two faces looking, simultaneously, into the future and the past. In fact quantum objects are usually even more enigmatic and have more than two faces, and sometimes infinitely many. When quantum objects let us look at their faces, though, they only show one face at a time: never both.

What we see is that repeated experiments with quantum objects aimed at measuring the value of some property will result in a series of different values in random order. Whenever a quantum object is queried by a measuring device, the inherently superposed quantum state (like a multifaceted Janus) collapses into one of the pure states and that is what we read on the dial of whatever contraption we use to do the experiment.

Repeating the experiment many times over will provide a number of different results in random order, but some values of the measured property will turn up more than others. In fact, the frequency of measuring any one particular value has a well-defined probability and can be predicted from theory with astonishing accuracy.

A fancy term used for the measurement problem is the collapse of the *wavefunction*. The wavefunction is the mathematical description of superposed states in Hilbert space invented by Erwin Schrödinger in his famous equation. The word 'wave' actually has a relevance to real waves on the surface of a lake. The collapse of the wavefunction can be visualized by looking at the rolling waves of a lake when suddenly, like magic, someone touches the surface of the water and makes the surface instantaneously become perfectly smooth—*except* for a single fountain of water that appears at a random location.

As time goes by the fountain slowly diminishes and fades into the reappearing rolling waves until another touch will cause the surface to smooth out again and another fountain springs to life. This really is everything we know about the measurement problem. A marvelous display of such fountain magic can be seen at Pierre du Pont's renowned Longwood Gardens.[9]

From our perspective it is the interpretation of the measurement problem that matters. The classical interpretation is the so-called Copenhagen interpretation and goes back to Niels Bohr's hub for Quantum revolutionaries in the 1920s and '30s. Niels Bohr was the man who just *knew* what is in the atom.

Once I took a long walk in the maze of narrow streets of old-town Copenhagen and came across a scene one finds in many old European cities, a town square filled with statues of kings, bishops, and generals of long forgotten wars. I casually scanned the faces of these historic figures and was astonished to realize that one of them looked eerily familiar. Well, there is no way I can possibly recognize any of these people so I walk up to the base of the statue to read the name of this gentleman and, you guessed it, he was Niels Bohr. His Copenhagen interpretation goes to the core of human psychology.

---

[9] See http://www.longwoodgardens.org.

When a psychologist queries the mind (sometimes under hypnosis) the information retrieved from the deep layers of the unconscious depends on how the doctor interacts with the patient. What this means is that the answer is necessarily influenced by the question itself: the psychologist's questions make the brain work to provide an answer. Unlike a star gazer who can look up at the starry sky and passively observe the constellations without ever influencing them, the psychologist is always actively engaging the patient's mind.

The core idea of Bohr's interpretation of the measurement problem is that the apparatus carrying out the measurement is always going to engage the quantum object to be measured; this disallows passive observation and means that the statistics of the outcome of a great many repeated measurements will ultimately be determined by the apparatus itself.

Granted; without a doubt this is bizarre for everyday thinking and, of course, the immediate question that comes to mind is, *WHY*? The short answer is simply a reminder that this is the Copenhagen *interpretation* and not the

Copenhagen explanation. Nonetheless if you ask proponents of the science of emergent phenomena they'll tell you that this is in fact, quite simple.

We don't know how to measure a quantum object one-on-one; that is to say, we do not know how to measure a single photon using another single photon.[10] The measuring apparatus is always a macroscopic contraption and as such is made of innumerable quantum objects similar to the one being measured. According to the law of emergence, the macroscopic contraption is more than the sum of its microscopic constituents; it is *more* in a way that is unique to that measuring instrument. The apparatus itself creates the laws governing the behavior of quantum objects participating in this interactive game we call the measurement process.

---

[10] In fact, as the 2012 Nobel Prize in Physics attests to it, the first steps have already been taken towards being able to manipulate *individual* quantum systems; in other words, to master quantum systems in their own right without bringing them into direct contact with the macroscopic world. So far these systems include single photons and also charged atoms called ions. For more information, refer to http://www.nobelprize.org/nobel_prizes/physics/laureates/2012/popular-physicsprize2012.pdf.

Hence the measurement process involves an object and an observer interacting in a way that, lacking a better understanding, can be likened to psychological inquiry. At any rate, the basic tenet of the Copenhagen interpretation is the coexistence of the *observed* and the *observer*. So far, the observed has been a single quantum object such as a photon, and the observer a macroscopic apparatus.

Of course, you would argue that the observer is really the person reading the dials on the measuring device, and *you are absolutely right*. It is the researcher's mind where the idea of a particular measurement takes shape in the first place. The actual contraption is a manifestation of that idea; the two cannot be separated. Moreover, the nature of what exactly is measured is immaterial from our perspective: the real question is *where to draw the boundary between observed and the observer*.

Why would researchers be limited to measuring a single quantum object? There already exist measurements where multiple quantum particles (even ordinary objects with tiny but macroscopic dimensions) are considered and it is observed how under certain circumstances these particles are entangled even though they are located too far apart to communicate.

This type of experiment proves that Albert Einstein's most famous criticism of quantum mechanics (often referred to as the EPR paradox[11]) was *wrong*. As astonishing and unbelievable it sounds, quantum systems can indeed expand to *macroscopic* dimensions yet obeying the laws of the *microscopic* world. These entanglement experiments are still in their infancy but their very existence hints at the heart of the measurement problem. Entanglement is an extreme manifestation of the superposition principle, the mystery of the Quantum that is so difficult to reconcile in the everyday world.

---

[11] Albert Einstein, Boris Podolsky, and Nathan Rosen, *Phys. Rev.* **47**, 1935, 777–780.

Entanglement is arguably a most exciting topic for lab tea discussions among scientists of all shapes and sizes but we shall resist the temptation to join in. It brings about physical evidence of the superposition principle at work in the macro world.

Just think about it—if superposition is at work in the macro world, the philosophical ramifications are astounding. Science's entrenched distinction between the micro and macro worlds is entirely arbitrary and so is the boundary between the *observed* and the *observer*.

If what is observed can be more than a single quantum object, it can be anything. For example, it can be the whole apparatus measuring the splitting of a single photon, to be measured in the famous double slit experiment where a single photon passes through two different gates at the same time; we just need a more sophisticated measuring device not yet invented. Or perhaps an advanced extraterrestrial civilization could visit Earth and take a measurement of the whole planet...or the Galaxy...or the Universe?

∞

The measurement problem has no limits: it applies to anything and everything. When it comes to the Universe, the measurement problem begs the most fundamental question that can be asked of a scientist:

WHO IS THE ULTIMATE OBSERVER?

# For in *Him* We Live

Who is the ultimate observer? There are only two mutually exclusive and equally momentous answers to this question. One includes a providential caring God, the other doesn't.

The answer that I have a deep conviction is the correct one is very simple and is inherent to Christian thought. The ultimate observer is the creator God who is both immanent and transcendent to the Universe. Being both immanent and transcendent is a divine attribute that forms the basis of *panentheism* found in many religions and philosophies. Unlike pantheism that essentially equates the Universe with God, panentheism asserts that the Universe is created by God as a temporal manifestation of divine thought. The Universe is not God, the Universe is *in God* and it is infused by God's love, grace, and providence.

Panentheism is at the heart of Arthur Peacocke's idea of whole-part-influence wherein the immanence of God is not just a passive imprint of His being in the world, but also an active way of exerting His providence *by Him being the ultimate observer*.

We set sail on the Quantum Ocean to come to this conclusion:

*Whole-part-influence is infused into the Universe in the depth of the Quantum, by God inquiring the world in perpetuity as the ultimate observer.*

This is what we meant when we said earlier that the world is breathing God. St. Paul tells the Athenians standing on the Areopagus:

*For in Him we live, and move, and have our being*
*(Acts 17:28)*

To have our being in God is the absolute submission of our self to His love, grace, and providence. We shall attempt to say more in the Epilogue, but the scientific inquiry has to stop here. As grand as science is, with its limited scope and means, it cannot grasp the full riches of providence.

∞

Nonetheless, it was pure scientific thought that conjectured the ultimate observer. Providence cannot be ruled out on scientific grounds. We can either recognize that the ultimate observer is God. Or we can deny it. Neither choice is scientific but both have repercussions beyond measure. I already stated my conviction, but with almost no exception main stream scientists will hear none of it. So, then, what is the *scientific* alternative to a providential God that captivates the minds of most of the scientific community?

In 1957, Hugh Everett, while a physics graduate student at Princeton University, published a ground-breaking article about the *relative state* formulation of quantum mechanics.[12] The interpretation of Everett's key idea of no-collapse wave mechanics has mind boggling implications leading up to the now popular *many worlds* view that is gaining warmer acceptance among physicists and is mesmerizing science fiction enthusiasts.

---

[12] Hugh Everett III, *Rev. Mod. Phys.* **29** (Nr. 3), 1957, 454-462.

Using the fountain-in-the-lake analogy, Everett's theory shows that, *mathematically*, the Bohr interpretation of an observer tapping the surface of the lake and causing the waves to instantaneously collapse into a single fountain is, astonishingly, equivalent to a no-collapse interpretation where all fountains spring to life simultaneously.

How is that possible, though, if we know from experiment that indeed the waves do collapse into a single fountain? The answer will split your mind, literally. Mathematics is one thing, reality is another. As long as only one fountain is real and the others are just mathematical ghost images we have a grip on reality. Have you ever been riding in one of those fancy mirror-walled hotel elevators where you can see an infinite number of reflections of yourself? One reality—you—and an endless succession of reflections. This is how Everett himself thought of his own theory and maintained that there was only *one* physical reality. [13]

---

[13] "Throughout all of a sequence of observation processes there is only one physical system representing the observer, yet there is no

The *many worlds* view, however, asserts (albeit without any foundation) that the world of mathematics is naturally real. The many worlds view suggests that our Universe is, in fact, a 'multiverse' in which a plethora of parallel realities co-exist independently. In the multiverse version of the mirror-walled elevator your reflections are distorted in a myriad different ways and represent realities of many different *You* living in independent, parallel worlds. The many worlds view argues that the measurement process involves an infinite succession of *spontaneous splitting* of the whole Universe into equally real worlds that collectively constitute the multiverse.[14]

---

single unique state of the observer (...). Nevertheless, there is a representation in terms of a superposition, each element of which contains a definite observer state and a corresponding system state." Hugh Everett III, *Rev. Mod. Phys.* **29** (Nr. 3), 1957, 454-462.

[14] In fancy physics lingo, simply put, the wavefunction of the Universe never collapses, instead a wavefunction describing one classical reality gradually evolves into a wavefunction describing the superposition of many such realities. Max Tegmark, *Nature* **448** (Issue Nr. 7149), 2007, 23-24.

With due respect, it puzzles me why creation via "spontaneous splitting" with no reason is more attractive to many scientists than creation via a flash of divine love and genius with a purpose, but I do accept that this fascinating theory is sound mathematically and its predictions are consistent with observation, which should suffice for scientific inquiry. Therefore, I ask two simple questions.

First, in Bohr's interpretation there is an observer tapping the surface of the lake with a purpose to create an awesome fountain spectacle. Note that the observer need not wait for a fountain to fade entirely into the rolling waves before tapping the water again. Doing so would be like playing the piano tapping the keys one by one, with a single index finger and not using the pedals. No one would recognize the music. The observer creates fountain magic like a piano genius creating music that is a concerto, not a cacophony, of many notes sounding together.

In the multiverse interpretation, however, there is no pianist. Instead, every once in a while the piano spontaneously sounds every single note while simultaneously splitting into a myriad of baby pianos. Yes, Everett's mathematics shows that people living in parallel universes can still hear music, but my question; my purely scientific question is *why does the piano split*? Why does the Universe split? Maybe there is something in the Universe building up and then boiling over and it just must split? I surely do not have the faintest idea. Nor does anyone else advocating the multiverse.

Then, consider that when repeating a quantum experiment many times, different outcomes occur with *well-defined probabilities*. The multiverse interpretation that every single outcome happens with a probability of one is not consistent with observed results of quantum experiments. Probabilities are supposed to add up to one, but in this case they would add up to $N$ where $N$ is the number of outcomes. $N$ is greater than or equal to 1 and it is not limited from above, $N$ can even be infinity.

$$\infty$$

Granted, Everett's mathematics comes to the rescue again, but this time with a caveat. It can be shown that people living in most (but not all) parallel universes measure probabilities indistinguishable from those measured in our Universe, but there would be some renegade universes where the probabilities would be different. An excellent short reading of the pro multiverse interpretation is Max Tegmark's commentary in the fiftieth anniversary issue of *Nature* on Everett's original paper.[15]

There could be counterarguments built around the anomaly of the renegade universes, but my question; my purely scientific question is simply *Why*? Why should all splits happen all the time? The multiverse interpretation has no answer.

Where is the matter coming from filling the newly born baby universes? Where is the energy coming from to make it all happen? Created out of nothing? Many more questions that do not have answers, pro or con, given the state of the art of our ideas about the world around us.

---

[15] Max Tegmark, *Nature* **448** (Issue Nr. 7149), 2007, 23-24.

There is another way, though, to approach this subject. I would like to put forward a different kind of argument regarding human nature. When it comes to the interpretation of multiverse theory, Everett's equations bear cardinal implications and if we accept the existence of parallel universes according to the many worlds view, we also have to accept far-reaching consequences, which heavily contradict our human nature and our science.

The many worlds view proclaims that each and every parallel world is one real face of a multifaceted reality and, eventually, anything and everything that can happen within the boundaries of the laws of physics will happen in one of the parallel universes. Therefore, the many worlds view *contradicts* free will as well as the idea of a unique person, being right or wrong, honor, human dignity, and the virtue of human efforts. In fact, the many worlds view also contradicts the wonders of evolution.

Free will has always been controversial, but it is hard to deny that free will has a legitimate place in human nature.

∞

Of course, we are wired by our genome, our heritage, and our social network, but a person can always say 'no' when everybody else says 'yes'. Some of history's most uplifting moments attest to this. Negating the idea of a unique person and moral values has already had dire consequences throughout history.

Denying human effort, denying *free will*, digs deep into the fabric of science itself. I would like to believe that in my own research field, the colossal human effort spent in trying to understand why protein molecules fold into one particular shape in the cell and not some other shape, has a legitimate value and will help cure diseases.

If, as a person and a scientist, I am forced to believe that somewhere some 'me' will help create heaven in some world and somewhere else some other 'me' will help create hell in another world, and the whole palette in between...then why bother?

The deep roots of the contradiction between the many worlds view and human nature lie in the equal footing of parallel universes. The many worlds view does not allow for a human making a *choice*. We humans arrive at crossroads all the time and make hard choices about which road we take to continue our journey. Once a choice has been made, often with agonizing pain, there is no way back; the alternative routes will never be explored. At least this is how undeniably most of us feel.

Of course, one could argue that human nature would also be different in parallel universes and, perhaps, in most other worlds people would not see any or much contradiction with their nature.

Notwithstanding that possibility, we do, however, see a contradiction with our own human nature. As beautiful a mathematical theory might be, it is subject to experimental verification according to the very nature of scientific thought. Aren't we humans also part of the experimental/observational picture? Can we, then, ignore our own self when it comes to accepting the many worlds view?

Some would argue that the most sensitive instrument is a pencil standing on its tip; the tiniest microscopic push would make it tip over. I would argue that the human person is even *more* sensitive, and used as an instrument to experimentally verify the multiverse theory, clearly *disproves* it and *refutes* the many worlds view.

Multiverse theory has come a long way in its mathematical formulation from Everett to the now popular M-theory, which is an advanced synthesis of string theories requiring eight extra spatial dimensions besides the familiar three, curled up in so tiny a scale that we may not be able to detect them.

In any case, no matter how grand and complex the mathematics might be, multiverse theory is seriously called into question by testing it experimentally on a human.

My simple conclusion is that Niels Bohr's Copenhagen interpretation that allows God to interact with our world in the depths of the Quantum as the Ultimate Observer is more sound scientifically than the multiverse interpretation.

Ultimately, the question is whether we go it alone or with God. Whether the Universe is in God and infused by God's love, grace, and providence; or the Universe (or multiverse) is entirely self-contained and has no purpose.

*"For in Him we live, and move, and have our being"*
*(Acts 17:28)*

St. Paul's quote from Scripture has solid scientific footing while the many worlds view trying to escape God is merely a mesmerizing and intricate sand castle.

∞

# Epilogue

Pierre Teilhard de Chardin, the renowned paleontologist and Jesuit priest, enshrined his revolutionary ideas on evolution in *The Phenomenon of Man*, paving the road from science toward God. His seminal work couldn't even be published until after his death in 1955 because it was considered too controversial for his time.

Chardin presented three ideas that laid the foundation for a new paradigm of evolution that goes beyond Darwinian natural selection and includes purpose. As you can imagine his ideas are still highly controversial but here they are.

1) Qualities that we consider as unique attributes to entities of a certain complexity are in fact present at levels of lesser complexity. Think of communication for example. Human speech is what we consider to be the most advanced communication. But what about the language that dolphins

speak, or birds, wolves, or even the mute fish? Consider the molecular level where intricate communication networks in biochemical pathways keep cell functions operational. Communication is not unique to only higher echelons of the tree of life and speech is not necessarily the highest level of communication. Humans can surely communicate without words. Life itself is perhaps the greatest mystery, where do we dare draw the line between life and not life?

2) Evolution has a clear direction in producing more and more complex nervous systems. In that respect, evolution never looks back. Once a solution has been found to make the nervous system more complex and more powerful, that solution is carried over to grow new branches on the tree of life. Why more complexity in the nervous system? For one thing, regarding the first Chardin idea, the nervous system evolves into the brain capable of thinking. Thinking may be dormant at lower levels, but it is surely manifest in the human brain and it would be

∞

foolish to draw the line between thinking and not thinking.

3) Regarding his first two ideas, Chardin set the scene of evolution on his Christian backdrop introducing his intellectual trademark legacy; the *Omega point*. He argues that if there is a direction in evolution there is also a destination. Of course, many would argue that complexity may just have no limits and extend to infinity without a destination or purpose. However, this is not what Chardin said. He identified the Omega point in Jesus Christ as the pinnacle of (human) evolution.

St. Paul called Jesus the second Adam, the man reborn. The second Adam is not some kind of superman: the second Adam is man whose relationship with God is restored. It is not his intellect, teaching, empathy, healing, nor his good deeds by themselves that make Jesus *Jesus*; it is all that but also his suffering and his unique relationship with his Father.

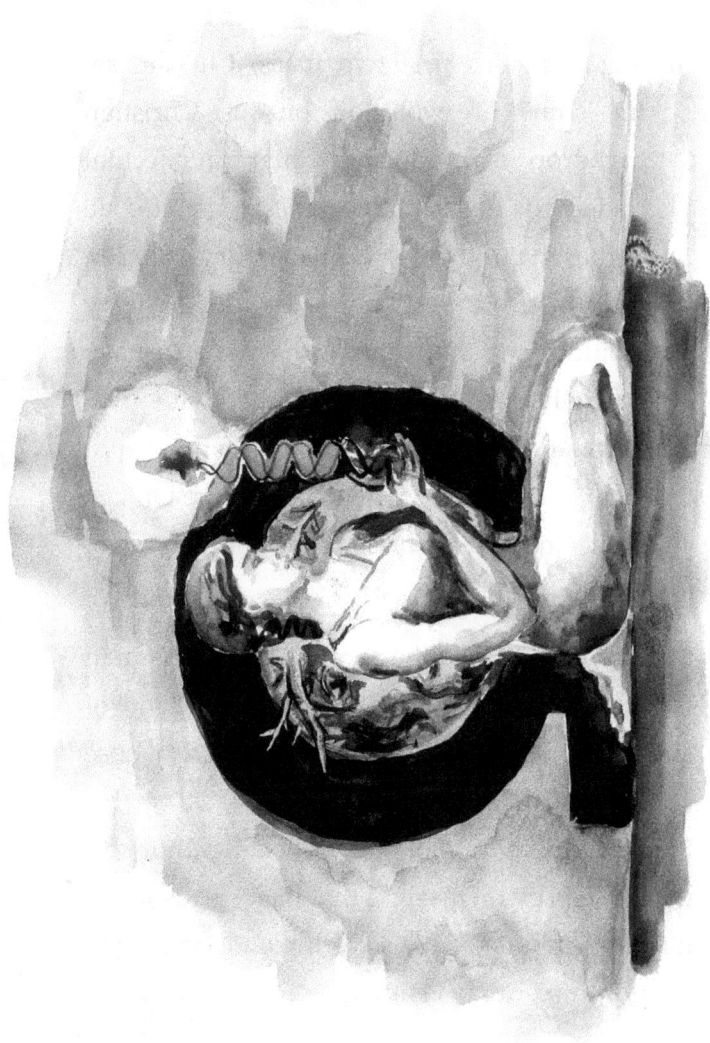

The Omega point is all about *relationship*; relationship with God and relationship with man. Chardin introduced the term *noosphere* representing, literally, a different kind of sphere beyond the layers of the atmosphere wrapping the Earth, glowing and radiating the collective human intellect into space and toward God.

This is very different from popular sci-fi themes of the Collective embodying absolute power over entire galactic species. The noosphere is where the unity of human intellect is manifest and where the collective human heart is in a love relationship with God. It has nothing to do with galactic powers.

The main body of my thesis is about simply arguing that one cannot exclude on scientific grounds that God could interact with the world, and I offer some thoughts how this could happen at the depths of the Quantum.

Interaction is the first—or more like the zeroth—prerequisite for relationship.

If I as a scientist give God an inch in the Quantum, He will want a foot, a yard, in fact the whole nine yards of me and will want a love relationship at every level of my being. You are no different, nobody is different. Give Him a tiny opening and your life will be enriched beyond measure. If there is anything in life without limits I believe it is God's resourcefulness to love people and teach them to love each other. This is the real meaning of providence, learning to love in relationship with God. The key is not to look outward or upward in search of God, look inside. Not only do we live in Him, He also lives within us.

I stated in the *Preamble* that I have no claim of being right or wrong. Do not think that I am telling you these things from experience. I wish I could. Like I said, I am on a pilgrimage and invited you to join me. That's all.

However, there is one more point I want to make before I can put the pen down. The Omega point, divine purpose and providence, and glowing noosphere sound so amazingly positive whereas when we look around we see a lot of dead ends, tragedies, violence and war, pain, poverty, hopelessness, and suffering. Really, we do look evil in the eye.

What did Chardin say about evil? Very little. In fact, *The Phenomenon of Man* is a 300-page book in which evil is dealt with in a 3-page Appendix and Chardin basically says that he is too optimistic to talk about it. He poses the question:

*"Throughout the long discussion we have been through, one point may perhaps have intrigued or even shocked the reader. Nowhere, if I am not mistaken, have pain or wrong been spoken of. Does that mean that, from the point of view I have adopted, evil and its problem have faded away and no longer count in the structure of the world?"*[16]

For his answer read the Appendix. There is a plethora of accounts on evil in writings, paintings, and films. How about *The Scream* by Edvard Munch, just to mention my favorite?

---

[16] Pierre Teilhard de Chardin, *The Phenomenon of Man*, Harper Perennial Modern Thought, New York, 2008, pp 311-314. Appendix, Some Remarks on the Place and Part of Evil in a World in Evolution.

With no further ado let me conclude with my personal account. When it comes to suffering, nothing is more compelling to me than the picture of Pope John Paul II in his final days clutching the cross kneeling in his private chamber watching the Good Friday procession at the Colosseum in Rome that he no longer could lead. His body ravaged by multiple death sentences, suffering in silent prayer: prayer that has nothing to do with trying to convey God to follow our will. When we ask God, it is like a little child saying I want this, I want that, not because I really want it or even know what it is that I want; it is a little child wanting attention. That is our first step toward God in prayer.

When I first saw this picture I knew his prayer was different, he wasn't the Pope any more, he was Karol Wojtyla not asking for anything, just silently co-suffering with Christ. He wrote a deep and sensitive encyclical letter called *Salvifici Doloris* in 1984 on the Christian meaning of human suffering.[17]

---

[17] Pope John Paul II, On the Christian Meaning of Human Suffering, Pauline Books and Media, Boston, 2009.

The following is my faint reflection on *Salvifici Doloris*. Feel free to ignore it, but I recommend you read *The Shack*.[18]

Human suffering is like love; one has to live it to let it sink into the depths of one's psyche. The meaning of suffering cannot be pieced together from its myriads of manifestations. We can reason about suffering to some degree but in the end the argument always feels hollow. We can argue, for example that Job's tragic story in the Old Testament makes it clear that the meaning of suffering is beyond the realm of justice. We can also argue that suffering can be the source of steel in our characters and furthermore it can teach us to appreciate things that we would otherwise take for granted.

This is all true but do you think that Mother Teresa had ever pondered these arguments in the slums of Calcutta? Can we hope to ever cope with and maybe someday do away with senseless human suffering? Maybe not: But does

---

[18] William P. Young, *The Shack*, Windblown Media, NewBury Park, CA, 2007.

that mean we cannot try? Does that mean we shouldn't try? Can we, should we, follow in the footsteps of Mother Teresa?

We must answer for ourselves. But remember one thing: Jesus Christ, Chardin's Omega point and the pinnacle of human evolution, died on the Cross amongst the horrors of human suffering. The Christian faith is built around Jesus Christ. Jesus Christ the person, Jesus Christ in his life's story. Jesus' life is Jesus Himself. We do not think of Jesus by his teaching or by his deeds, we do not reference his works. When we think of Jesus it is of his life.

One of the highlights of a New York Christmas is the annual Radio City Music Hall Christmas Spectacular. It is a great show with the astounding Rockettes. But in the finale the curtain falls over a live Nativity scene and the story of *One Solitary Life* by James Allen Francis is told in giant scrolling text about a man who never wrote a book, never held an office, never went to college, never visited a big city, never traveled more than two hundred miles from the place where he was born, did none of the things usually associated with greatness, and had no credentials but himself.

Yet all the armies that have ever marched, all the navies that have ever sailed, all the parliaments that have ever sat, all the kings that ever reigned put together have not affected the life of mankind on earth as powerfully as that one solitary life.

The central teaching of Christianity simply says follow Jesus and by doing that we may realize that evil can only be eradicated by suffering. The Christian meaning of human suffering is, paradoxically; eradicating evil.

Amazingly, *Salvifici Doloris'* deep spiritual insight is quite rational and gives the recipe for every person to follow Jesus whether one is Christian or not. The simple *equation* is that the suffering of a single person should awake empathy in a whole community. Suffering is individual but the good Samaritan should be a community rather than a single good soul.

More suffering should awake even more empathy and, through love, bring about action to put an end to suffering. This is the only way.

*'Come, you who are blessed by my Father;*
*take your inheritance, the kingdom prepared*
*for you since the creation of the world.*
*For I was hungry and you gave me something to eat,*
*I was thirsty and you gave me something to drink,*
*I was a stranger and you invited me in,*
*I needed clothes and you clothed me,*
*I was sick and you looked after me,*
*I was in prison and you came to visit me.'*

...

*'Truly I tell you, whatever you did for one of the least*
*of these brothers and sisters of mine, you did for me.'*

*(Matthew 25:34-36, 40)*

∞